NANOTECHNOLOGY

EXPLORING THE SMALL WONDERS OF THE FUTURE

CYRIL LAKES

Contents

- CHAPTER ONE 3
- Introduction 3
  - Principles of Nanotechnology 17
- CHAPTER TWO 19
  - Nanomaterials and nanoparticles 23
  - Nanotechnology's Environmental Applications 42
- CHAPTER THREE 46
  - Implications for society and ethics...........................56
  - Public perception and education 63
- CHAPTER FOUR 67
  - Prospects for the future and upcoming developments 69
  - Conclusion..76
- THE END 80

CHAPTER ONE

Introduction

Nanotechnology is a multidisciplinary study that focuses on manipulating matter at the atomic, molecular, and supramolecular scale. It is often referred to as the science of the small. At this scale, materials display distinct properties and behaviors that are different from those observed at larger scales, providing possibilities for innovation and applications in several areas. Nanotechnology possesses the capacity to fundamentally transform various domains like medical, electronics, energy, and materials science. This has far-reaching consequences for a wide range of applications, spanning from healthcare and environmental restoration to computing and manufacturing.

The term "nano" is a unit of measurement equal to one billionth of a meter, or 10^{-9} meters, emphasizing the extremely minute scale at which nanotechnology functions. At this level of measurement, phenomena controlled by quantum mechanics become more and more common, resulting in new characteristics like quantum confinement, surface effects, and increased responsiveness. Scientists in the field of nanotechnology utilize these qualities to create and manipulate materials, structures, and devices with specific functions and performance characteristics.

Nanotechnology involves a diverse array of methods and strategies, such as bottom-up synthesis, top-down production, self-assembly, and manipulation at the atomic and molecular scale. Scientists employ techniques like scanning probe microscopy, electron microscopy, and molecular modeling to observe, analyze, and control structures and materials at the nanoscale.

Nanotechnology finds applications in various sectors, such as:

Nanotechnology has potential for progress in medication delivery, diagnostics, imaging, and therapies within the field of medicine and healthcare. Nanoscale drug delivery systems possess the ability to selectively target particular cells or tissues, thereby augmenting the effectiveness of drugs while simultaneously reducing the occurrence of undesirable side effects. Nanomaterials utilized in diagnostics and imaging facilitate the prompt identification of diseases and offer valuable understanding of biological mechanisms.

Nanotechnology is fueling advancements in electronics, resulting in the development of smaller, quicker, and more energy-efficient gadgets. The utilization of nanoscale transistors, memory devices, and sensors facilitates the advancement of electronics and computer technologies of the future. Nanomaterials such as quantum dots, nanowires, and nanotubes have various applications in the field of electronics.

Nanotechnology provides solutions for the generation, storage, and conversion of clean energy in the field of energy and environment. Nanomaterials, such as solar cells, fuel cells, and batteries, facilitate the effective acquisition and retention of energy. Nanotechnology aids in environmental remediation by creating filters, catalysts, and sensors using nanomaterials to regulate pollutants and purify water.

Nanotechnology in Materials Science and Engineering allows for the creation and production of sophisticated materials that have specific features and functions. Nanocomposites, nanocoatings, and nanofibers possess improved mechanical, thermal, and electrical characteristics, making them suitable for use in aerospace, automotive, construction, and consumer goods industries.

Nanotechnology allows for exact manipulation and organization of nanoscale structures and devices in the manufacturing and nanofabrication processes. Methods such as nanoimprint lithography, electron beam lithography, and molecular self-assembly allow for precise and detailed shaping and control of materials at the nanoscale. This opens up possibilities for the advancement of nanoelectronics, nanophotonics, and nanobiotechnology.

Nanotechnology's ongoing progress offers the opportunity to tackle critical societal concerns, encompassing healthcare, energy, sustainability, and global connectivity. Nevertheless, nanotechnology also presents significant ethical, safety, and regulatory concerns that need to be resolved in order to guarantee responsible and sustainable progress. By utilizing the potential of nanotechnology and effectively tackling its obstacles, scientists and pioneers might discover fresh possibilities for revolutionary solutions that enhance the standard of living and propel advancements in the 21st century and beyond.

Nanotechnology refers to the manipulation and control of matter at the nanoscale, which is the scale of atoms and molecules. It involves the design, synthesis, and application of materials and devices with unique properties and functionalities at this scale. The scope of nanotechnology encompasses various fields, including physics, chemistry, biology,

Nanotechnology is an interdisciplinary domain that encompasses the manipulation, characterisation, and utilization of materials and structures at the nanoscale, typically measuring between 1 to 100 nanometers (nm) in size. At this particular size, materials display distinct physical, chemical, and biological characteristics that are different from their larger versions. Nanotechnology is a field that covers many different scientific areas, such as physics, chemistry, biology, materials science, engineering, and computer science. It is used in several industries, including healthcare, electronics, energy, and environmental remediation.

The domain of nanotechnology encompasses:

Nanotechnology utilizes both bottom-up and top-down methodologies to construct tiny structures and gadgets. Bottom-up methods entail the construction of nanoscale structures by manipulating individual atoms or molecules using techniques including self-assembly, chemical synthesis, and molecular manipulation. Top-down approaches encompass the utilization of lithography, milling, and deposition processes to sculpt or engrave bulk materials, resulting in the creation of nanoscale characteristics.

Nanotechnology is a field that specifically deals with the creation, production, analysis, and usage of nanomaterials. Nanomaterials are materials that have at least one dimension that falls within the nanoscale range. Nanomaterials possess distinctive characteristics, including a high ratio of surface area to volume, quantum confinement, and adjustable optical, electrical, and mechanical properties. Nanomaterials encompass a variety of substances, such as nanoparticles, nanotubes, nanowires, quantum dots, and nanocomposites.

Nanotechnology focuses on the creation and advancement of technologies and systems at the nanoscale level, with the aim of utilizing them in diverse applications. Nanodevices consist of several technologies, such as nanoelectronics, nanophotonics, nanomechanics, and nanosensors. Nanosystems combine several nanodevices to create operational systems designed for specific purposes, such as medical diagnostics, environmental monitoring, and information processing.

Nanotechnology utilizes sophisticated characterisation tools to see, control, and examine structures and materials at the nanoscale. Scanning probe microscopy, transmission electron microscopy, atomic force microscopy, and X-ray diffraction are methods that allow researchers to investigate the characteristics, composition, and behavior of nanomaterials at the atomic and molecular scale.

Applications in Multiple areas: Nanotechnology is widely used in different areas, encompassing:

Nanotechnology revolutionizes healthcare by facilitating progress in drug administration, medical testing, visualizing internal structures, and restoring damaged tissues. Its potential uses encompass cancer therapy, tailored medical treatments, and the creation of artificial organs.

Nanotechnology revolutionizes electronics, photonics, and optoelectronics, resulting in the development of smaller, quicker, and more energy-efficient devices including transistors, screens, and sensors.

Nanotechnology provides solutions for the production, storage, and transformation of renewable energy, specifically in the areas of solar cells, batteries, fuel cells, and catalysis for environmental cleanup.

Materials Science and Engineering: Nanotechnology allows for the creation of sophisticated materials that have specific qualities and functions. These materials include nanocomposites, nanocoatings, and nanofibers, which can be used in various industries such as aerospace, automotive, construction, and consumer items.

Nanotechnology plays a significant role in the progress of information technology, namely in the areas of data storage, quantum computing, and nanophotonics, which enable high-speed communication and processing.

The field of nanotechnology is making tremendous progress due to continuing global research and development endeavors. As the area advances, it has the capacity to tackle the most urgent concerns confronting society, encompassing healthcare, energy, sustainability, and global connectivity. Nevertheless, nanotechnology also presents significant ethical, safety, and regulatory concerns that need to be resolved in order to guarantee responsible and sustainable progress. By utilizing the potential of nanotechnology and overcoming its obstacles, scientists and inventors might discover fresh possibilities for revolutionary solutions that enhance quality of life and propel advancement in the 21st century and beyond.

Principles of Nanotechnology

Nanotechnology fundamentals involve the fundamental principles, concepts, and techniques used to manipulate, characterize, and apply materials and structures at the nanoscale. Comprehending these basic principles is essential for progressing nanotechnology research and creating groundbreaking applications in diverse fields. Here are some essential principles of nanotechnology:

Nanotechnology focuses on materials and structures that are within the range of 1 to 100 nanometers (nm) in size, which is considered the nanoscale. At this particular magnitude, materials display distinct characteristics and actions that are distinct from those found at bigger magnitudes. Understanding the importance of nanoscale dimensions and the impact of quantum mechanics and surface phenomena on material characteristics is crucial.

Quantum mechanics is the branch of physics that describes the behavior of particles at the nanoscale and has a profound impact on the characteristics of nanomaterials. Understanding the behavior of electrons, photons, and other particles in nanoscale systems requires a grasp of important concepts such as quantum confinement, tunneling, and wave-particle duality.

CHAPTER TWO

Nanotechnology is concerned with the creation, production, analysis, and usage of nanomaterials, which are materials that have at least one dimension in the nanoscale range. Nanomaterials encompass a variety of substances such as nanoparticles, nanotubes, nanowires, quantum dots, and nanocomposites. Every category of nanomaterial has distinct characteristics and capabilities that make them well-suited for particular uses.

Nanotechnology utilizes a range of methods to synthesize and fabricate nanomaterials and nanostructures. Bottom-up procedures entail the construction of nanoscale structures by assembling individual atoms or molecules using techniques such as self-assembly, chemical vapor deposition, and molecular beam epitaxy. Top-down approaches utilize methods like as lithography, milling, and etching to carve or etch bulk materials in order to create nanoscale features.

Characterization Techniques: Advanced techniques for characterizing nanoscale structures and materials are crucial for their visualization, manipulation, and analysis. Scanning probe microscopy techniques, such as atomic force microscopy and scanning tunneling microscopy, along with transmission electron microscopy, X-ray diffraction, and spectroscopy, allow researchers to investigate the atomic and molecular properties, composition, and behavior of nanomaterials.

Nanotechnology utilizes the distinctive characteristics of nanoparticles to fulfill a diverse array of purposes in multiple industries. Nanomaterials demonstrate improved mechanical, thermal, electrical, and optical characteristics as a result of their diminutive dimensions and elevated surface area-to-volume ratio. Nanotechnology finds applications in various fields such as healthcare (for drug delivery and diagnostics), electronics (for nanoelectronics and sensors), energy (for solar cells and batteries), materials science (for nanocomposites and coatings), and environmental remediation.

Safety and Ethics: With the progress of nanotechnology, it is crucial to contemplate the possible environmental, health, and safety consequences of nanomaterials and nanostructures. Gaining knowledge on the toxicity, biocompatibility, and environmental consequences of nanoparticles is essential for the appropriate advancement and use of these materials. Attention should also be given to the ethical considerations regarding privacy, security, and societal ramifications of nanotechnology.

By acquiring expertise in these basic principles, scientists and engineers can make progress in the area of nanotechnology and create groundbreaking solutions to tackle worldwide problems, all while guaranteeing the conscientious and ethical application of nanomaterials and nanotechnologies.

Nanomaterials and nanoparticles

Nanomaterials and nanoparticles play a crucial role in the field of nanotechnology, as they possess distinctive characteristics and capabilities resulting from their minute dimensions and large surface area-to-volume ratio. Gaining knowledge about the properties and uses of nanoparticles is essential for effectively utilizing their capabilities in several industries. Below is a concise summary of nanomaterials and nanoparticles:

Nanomaterials are substances that have at least one dimension within the range of 1 to 100 nanometers (nm). Nanoparticles, nanowires, nanotubes, nanosheets, and nanostructures are among the different shapes in which they can exist.

Nanomaterials possess distinct properties and behaviors that deviate from those of larger materials as a result of quantum confinement, surface effects, and other nanoscale phenomena. The features encompass improved mechanical strength, electrical conductivity, thermal stability, optical transparency, and chemical reactivity. Nanomaterials can exhibit size-dependent features, meaning that their qualities vary depending on their size.

Synthesis: Nanomaterials can be produced by either bottom-up or top-down methods. Bottom-up procedures entail the construction of nanoscale structures by assembling individual atoms or molecules using techniques like self-assembly, chemical vapor deposition, and sol-gel synthesis. Top-down methods utilize lithography, milling, and laser ablation to carve or etch bulk materials, creating nanoscale features.

Nanoparticles are a specific kind of nanomaterial that is distinguished by its diminutive size and its significant surface area-to-volume ratio. They can consist of a variety of materials, such as metals, metal oxides, semiconductors, polymers, and carbon-based compounds. Nanoparticles possess distinct optical, electrical, magnetic, and catalytic characteristics, rendering them highly attractive for utilization in fields such as electronics, medicine, catalysis, sensing, and environmental remediation.

Purposes:

Healthcare: Nanoparticles are utilized in drug delivery systems to enhance drug solubility, stability, and targeting. Nanoparticles have the ability to enclose medications, genes, or imaging agents and transport them to particular cells or tissues, reducing unwanted effects and improving the effectiveness of treatment.

Electronics: Nanoparticles are integrated into electrical devices, such as transistors, screens, and sensors, in order to improve their performance and functionality. Quantum dots are semiconductor nanoparticles that possess adjustable optical properties, making them valuable for applications such as displays, lighting, and photovoltaic devices.

Nanoparticles operate as catalysts in chemical reactions because of their large surface area and reactive surface locations. They enhance the effectiveness and specificity of catalytic processes in several applications, including fuel production, environmental cleanup, and chemical synthesis.

Nanoparticles are utilized in environmental remediation for tasks like purifying water, filtering air, and remediating soil. They have the ability to extract pollutants, heavy metals, and toxins from environmental matrices, hence aiding in pollution management and cleanup endeavors.

Nanomaterials have various advantages, but their distinct characteristics give rise to apprehensions regarding potential hazards to the environment, health, and safety. Ongoing research is being conducted to investigate the toxicity, biocompatibility, and environmental impact of nanomaterials. Regulatory bodies around the world are in the process of drafting rules and regulations to assure the safe handling, use, and disposal of nanomaterials.

In general, nanomaterials and nanoparticles are extremely important in the progress of nanotechnology and the promotion of innovation in different industries. Researchers and engineers may exploit the potential of nanomaterials to create innovative technologies and solutions for various difficulties by comprehending their properties, synthesis methods, and applications.

The application of nanotechnology in the field of medicine and healthcare.

Nanotechnology has transformed the domain of medicine and healthcare by facilitating novel methods for diagnosis, therapy, drug administration, imaging, and monitoring. Nanomaterials and nanodevices provide distinct characteristics and capabilities that have facilitated progress in tailored medicine, focused treatments, and less intrusive procedures.

Nanotechnology is revolutionizing the field of medicine and healthcare.

Nanotechnology facilitates precise and regulated medication delivery systems, enhancing the effectiveness and safety of therapeutic treatments. Nanoparticles, liposomes, dendrimers, and polymer-based nanocarriers have the ability to enclose medications, genes, or imaging agents and transport them to particular cells, tissues, or organs. Targeted medication delivery reduces the occurrence of adverse effects throughout the body, improves the effectiveness of treatment, and allows for the gradual release of pharmaceuticals over long periods of time.

Nanotechnology provides hopeful possibilities for the diagnosis and treatment of cancer. Nanoparticle-based imaging agents facilitate the early identification of cancers using non-invasive imaging methods as magnetic resonance imaging (MRI), computed tomography (CT), and positron emission tomography (PET). In addition, nanoparticles have the ability to transport chemotherapeutic medicines, photothermal agents, or radioisotopes directly to cancer cells. This improves the effectiveness of treatment while reducing harm to healthy tissues.

Nanotechnology is essential in the field of regenerative medicine and tissue engineering. Nanomaterials, including scaffolds, hydrogels, and nanofibers, imitate the extracellular matrix and offer structural reinforcement for the growth, differentiation, and regeneration of cells and tissues. Nanoparticles containing growth factors, cytokines, or stem cells have the ability to stimulate tissue repair and regeneration in damaged or diseased tissues. This presents promising therapeutic options for disorders like spinal cord injury, heart disease, and osteoarthritis.

Nanotechnology improves the sensitivity, specificity, and resolution of diagnostic imaging techniques employed in medical imaging. Nanoparticle-based contrast agents, quantum dots, and magnetic nanoparticles allow for precise imaging of biological structures and processes at the molecular and cellular scale, achieving great resolution. These imaging agents have the ability to selectively bind to certain biomarkers, identify disease biomarkers, and continuously track the advancement of diseases in real-time. This enhances the ability to detect and diagnose diseases including cancer, cardiovascular disease, and neurological problems at an early stage.

Theranostics is a field that combines therapy and diagnostics using nanotechnology to provide personalized medicine strategies that merge diagnosis and treatment into a unified platform. Nanoparticle-based theranostic agents have the ability to deliver therapeutic medications and imaging agents to sick tissues at the same time. This enables the real-time monitoring of treatment response and the optimization of therapy regimens based on specific patient characteristics.

Nanotechnology facilitates minimally invasive surgical methods, offering improved precision, safety, and effectiveness. Surgeons can achieve precise interventions at the cellular or molecular level by using nanoscale surgical instruments, imaging-guided nanosurgery, and targeted medication delivery systems. This approach minimizes tissue damage and shortens recovery time. Nanoparticle-based contrast agents and imaging probes enhance the ability to visualize and navigate during surgical procedures, leading to better surgical outcomes and increased patient safety.

Nanotechnology provides options for controlling infectious diseases through the creation of antimicrobial nanoparticles, nanostructured surfaces, and nanofiltration membranes. Nanoparticles, which have been covered with antimicrobial substances, can impede the proliferation of bacteria, viruses, and fungi, hence diminishing the likelihood of healthcare-related infections and drug-resistant microorganisms. Nanostructured surfaces and coatings have the ability to inhibit the formation of biofilms on medical devices, implants, and hospital surfaces, thereby enhancing infection control and ensuring patient safety.

nanotechnology has great potential for enhancing the field of medicine and healthcare by providing groundbreaking solutions for the diagnosis, treatment, and prevention of diseases. Through the utilization of the distinct characteristics of nanomaterials and nanodevices, scientists and medical professionals have the ability to create customized treatments, enhance patient results, and tackle critical healthcare obstacles. Nevertheless, further investigation is necessary to tackle the safety, regulatory, and ethical concerns linked to the utilization of nanotechnology in the field of medicine and healthcare.

The application of nanotechnology in the fields of electronics and computing.

The application of nanotechnology has significantly influenced the domains of electronics and computers, facilitating progress in the reduction of device size, enhancement of performance, and improvement of energy utilization. Here are some primary methods via which nanotechnology is employed in various areas:

Nanotechnology enables the production of electronic components and devices at a very small scale, resulting in a significant reduction in size without compromising or even improving their performance. As a result, there has been a proliferation of compact and high-performance electronic gadgets, such as smartphones, laptops, and wearables.

Transistors, the basic unit of contemporary electronics, have tremendously profited from nanotechnology. As the size of transistors decreases to the nanoscale, their performance enhances, allowing for faster and more energy-efficient functioning. Nanotechnology concepts have been utilized to build advanced technologies like FinFETs (Fin Field-Effect Transistors) and nanowire transistors.

Nanotechnology has facilitated the creation of high-capacity memory technologies, including NAND flash memory and MRAM (Magnetoresistive Random Access Memory). These devices employ nanoscale architecture for data storage and retrieval, resulting in enhanced storage capacity and quicker access times.

Nanomaterials, such as carbon nanotubes and graphene, have enabled the development of flexible electronics. These materials exhibit extraordinary mechanical characteristics and can be incorporated into pliable bases, facilitating the creation of flexible screens, wearable sensors, and electronic skin.

Nanotechnology is essential for the advancement of quantum computing systems. Quantum dots, superconducting qubits, and other nanoscale structures are employed to generate and control quantum bits (qubits), which are the fundamental components of quantum information processing. These innovations have the potential to significantly increase computational speed relative to traditional computers for specific problem types.

Nanotechnology has enabled the creation of highly efficient systems for harvesting and storing energy. Nanomaterials, such as nanostructured silicon and quantum dots, are employed in solar cells to augment the absorption of light and improve the efficiency of conversion. In addition, researchers are investigating nanoscale materials for their potential use in high-capacity batteries and supercapacitors, which could offer quicker charging rates and longer lifetimes.

Nanotechnology allows for the creation of exceptionally responsive sensors and detectors that can be utilized in a wide range of applications. Nanoscale structures can be designed to detect extremely small variations in temperature, pressure, chemical composition, or biological molecules, rendering them highly desirable for applications such as environmental monitoring, medical diagnostics, and security systems.

Thermal management: The dissipation of heat poses a major obstacle in electronic devices. Nanotechnology-enabled thermal interface materials, such as films made of carbon nanotubes and polymers with nanocomposites, provide enhanced thermal conductivity, hence enhancing heat dissipation and increasing the reliability of devices.

nanotechnology remains a driving force behind advancements in electronics and computers, expanding the limits of device performance, energy efficiency, and functionality. As advancements in nanoscience research continue, we can anticipate other significant discoveries that will greatly influence the future of technology in these specific areas.

Nanotechnology's Environmental Applications

Nanotechnology has a broad spectrum of uses in the field of environmental science and engineering, offering creative methods for preventing pollution, cleaning up contaminated areas, monitoring environmental conditions, and conserving resources. Below are few significant environmental uses of nanotechnology:

Nanotechnology-based filtration systems, utilizing membranes covered with nanoparticles or nanotubes, are highly efficient in eliminating various water contaminants, such as heavy metals, organic pollutants, and pathogens. Nanomaterials possessing elevated surface area and reactivity augment the efficacy of adsorption, catalysis, and separation processes employed in water treatment.

Air filtration involves the use of nanofiber filters made from nanostructured materials such as carbon nanotubes or graphene oxide. These filters are designed to effectively trap particulate debris, allergens, and hazardous gasses found in both indoor and outdoor air. These filters provide excellent effectiveness and minimal pressure decrease, enhancing the quality of indoor air and diminishing the chances of respiratory health hazards.

Nanoremediation is a method that utilizes nanoparticles to break down, immobilize, or isolate harmful substances in the soil and water in contaminated sites. Nanomaterials, including zero-valent iron nanoparticles, titanium dioxide nanoparticles, and carbon nanotubes, can enhance the breakdown of organic pollutants, the containment of toxic metals, and the restoration of areas affected by oil spills and industrial waste disposal.

Nanotechnology provides prospects for enhancing agricultural methods and reducing environmental consequences in sustainable agriculture. Nanopesticides and nanofertilizers, which utilize nanoencapsulation or nanoparticle delivery methods, facilitate precise administration of agrochemicals, thereby minimizing chemical leaching, runoff, and environmental pollution. Nanosensors and nanoscale imaging techniques aid in precision agriculture by improving resource utilization and reducing pollution.

Nanoscale sensors and detectors allow for the continuous and immediate monitoring of environmental factors such as air and water quality, soil condition, and biodiversity. Nanomaterial-based sensors have the capability to detect extremely low levels of pollutants, diseases, and dangerous substances.

CHAPTER THREE

This enables them to serve as early warning systems for environmental risks and allows for prompt actions.

Nanotechnology enhances waste management by introducing advancements in waste reduction, recycling, and resource recovery, leading to more efficient procedures. Nanomaterials are employed in the creation of lightweight and long-lasting packaging materials, effective catalytic converters for waste-to-energy conversion, and enhanced separation methods for recycling electronic trash and other materials.

Nanotechnology improves the effectiveness and durability of renewable energy systems, including solar cells, fuel cells, and energy storage devices. Nanomaterials such as quantum dots, nanowires, and perovskite structures enhance the ability of solar panels, fuel electrodes, and batteries to absorb light, carry charges, and facilitate catalytic reactions. This leads to increased energy conversion efficiencies and extended device lifetimes.

Nanotechnology-based solutions have a role in mitigating climate change by decreasing the release of greenhouse gases, promoting energy efficiency, and improving carbon capture and storage technologies. Nanomaterials are used in catalytic converters to decrease automobile emissions, in energy-efficient building materials to decrease the need for heating and cooling, and in carbon capture membranes and adsorbents to capture and store CO2 from industrial flue gases.

These examples illustrate how nanotechnology may effectively and economically handle environmental concerns and promote sustainable development by offering efficient and environmentally friendly solutions in many sectors. Nevertheless, it is crucial to take into account the possible hazards and ethical ramifications linked to the use of nanomaterials in the environment, guaranteeing the responsible and secure implementation of nanotechnology-based solutions.

The application of nanotechnology in the field of manufacturing and industry.

Nanotechnology has transformed manufacturing and industry by allowing for the precise creation, construction, and control of materials and structures at the nanoscale. As a result, there has been a rise in the creation of sophisticated materials, procedures, and goods that offer improved performance, usefulness, and efficiency. Below are few significant applications of nanotechnology in the manufacturing and industry sectors:

Nanotechnology enables the creation of new materials with specific features and functions. Nanocomposites consist of nanoscale reinforcements that are evenly distributed within a matrix material. These materials demonstrate enhanced mechanical, thermal, and electrical characteristics as compared to traditional materials. Some examples of advanced materials are polymers reinforced with carbon nanotubes, composites made with graphene, and metals and ceramics with nanostructured properties.

Nanomanufacturing Processes: Nanotechnology facilitates meticulous regulation and manipulation of materials and structures at the nanoscale, resulting in the advancement of nanomanufacturing techniques. These approaches encompass top-down techniques like lithography, etching, and nanoimprinting, as well as bottom-up techniques like self-assembly, chemical vapor deposition, and atomic layer deposition. These procedures are employed to manufacture components, electronics, and surfaces at the nanoscale level with exceptional accuracy and consistency.

Nanotechnology is essential in the semiconductor business since it allows for the ongoing reduction in size of electrical and photonic devices, thereby playing a vital role in microelectronics and photonics. Advanced lithography and deposition techniques are used to build nanoscale components, including transistors, interconnects, and memory cells. Nanomaterials, such as quantum dots and nanowires, are utilized in electrical and optoelectronic devices to augment their performance and functionality.

Nanotechnology in Energy: Nanomaterials and nanodevices are applied in different energy-related applications, including solar cells, fuel cells, batteries, and energy storage systems. Nanostructured materials, such as perovskite solar cells and nanocomposite electrodes, improve energy conversion and storage efficiency. Nanotechnology facilitates the creation of energy harvesting and storage devices that are both lightweight and flexible, making them suitable for use in portable electronics and renewable energy systems.

Biomedical Manufacturing: Nanotechnology has altered biomedical manufacturing by enabling the manufacture of nanoscale drug delivery systems, diagnostic gadgets, and tissue engineering scaffolds. Nanoparticles and nanocarriers are utilized to transport therapeutic substances to particular sites in the body, enhancing treatment efficacy and decreasing side effects.

Nanomaterials are utilized in biosensors, imaging agents, and regenerative medicine applications.

Nanotechnology improves precision, efficiency, and sustainability in existing manufacturing technologies and processes, hence enhancing advanced manufacturing technologies. Industrial components and goods can be enhanced by nanoscale coatings and surface treatments, which offer benefits such as wear resistance, corrosion protection, and antibacterial qualities. 3D printing and other additive manufacturing methods can employ nanoparticles to create intricate structures that have excellent resolution and mechanical durability.

Nanotechnology plays a crucial role in promoting environmental and sustainable manufacturing through its ability to minimize waste, energy usage, and overall environmental harm. Nanomaterials are utilized in catalysts, filtration membranes, and adsorbents to mitigate pollution, cleanse wastewater, and purify air. Nanotechnology sensors and monitoring systems also enhance process optimization and resource efficiency in manufacturing operations.

Nanotechnology is employed in aerospace and defense sectors to make use of nanomaterials and nanodevices for the purposes of reducing weight, strengthening structures, and improving overall performance. Carbon nanotubes, graphene, and nanocomposites are used in aircraft components, armor materials, and electronics to enhance the ratio of strength to weight, increase durability, and improve functionality.

nanotechnology has unique prospects for creativity and progress in manufacturing and industry across several fields, promoting economic expansion, competitiveness, and sustainability. Nevertheless, it is crucial to acknowledge and tackle the possible hazards and difficulties linked to nanomaterials, including their effects on health and the environment, adherence to regulations, and ethical concerns. This is necessary to guarantee a responsible and secure implementation of nanotechnology in the manufacturing and industrial sectors.

Implications for society and ethics

The swift advancement and extensive acceptance of nanotechnology give rise to numerous societal and ethical ramifications that necessitate meticulous deliberation. These are the main areas of concern:

Nanomaterials have the potential to cause health hazards because of their distinctive physicochemical features. Exposure to nanoparticles through inhalation, ingestion, or contact with the skin can result in negative health consequences, such as inflammation, oxidative stress, and harm to cells. Extensive research is necessary to comprehend the enduring health consequences of being exposed to nanomaterials, and it is imperative to enforce suitable safety precautions in both work environments and consumer goods.

Environmental Impact: The introduction of nanomaterials into the environment during the process of manufacturing, utilization, and disposal presents ecological hazards. Nanoparticles have the potential to amass in soil, water, and ecosystems, which can have an impact on biodiversity, the functioning of ecosystems, and the food chains within them. Environmental surveillance and evaluation are crucial in order to minimize any environmental risks linked to nanotechnology.

Nanotechnology facilitates the creation of exceptionally perceptive sensors and surveillance technologies, which give rise to apprehensions regarding privacy and human liberties. Pervasive surveillance of individuals' activities, behaviors, and physiological data using nanoscale sensors could result in potential misuse and violation of privacy rights. In order to guarantee transparency, accountability, and safeguard personal data in the implementation of surveillance systems based on nanotechnology, it is imperative to establish strong legal and ethical frameworks.

The uneven allocation of advantages and disadvantages linked to nanotechnology might worsen pre-existing societal gaps and inequality.

The availability of nanotechnology-based products, services, and healthcare interventions may be restricted due to socioeconomic circumstances, geographical location, and regulatory obstacles, hence exacerbating the disparity between rich and underprivileged communities. It is crucial to make deliberate efforts to ensure fairness, availability, and participation for all in the advancement and implementation of nanotechnology in order to address issues related to social justice.

The ethical use of nanotechnology involves considering its appropriate and sustainable application in tackling societal concerns. Ethical considerations encompass the possibility of nanoparticles being misused for military applications, surveillance, and biowarfare, along with worries over unintended repercussions and unforeseen risks. Establishing ethical principles and codes of conduct is crucial for fostering ethical research, innovation, and governance in the field of nanotechnology. These guidelines and codes ensure that nanotechnology is applied in a responsible and constructive manner for the betterment of society.

The swift progress of nanoscale innovation presents difficulties for regulatory bodies in evaluating and controlling possible hazards to human health, safety, and the environment. The current regulatory frameworks may not be sufficient to effectively handle the distinctive characteristics and behaviors of nanomaterials, resulting in deficiencies in supervision and control. Enhancing regulatory capability and fostering global cooperation are essential to guarantee the secure and ethical advancement and commercialization of nanotechnology products and applications.

Public engagement and awareness play a crucial role in shaping the acceptability, adoption, and governance of nanotechnology, as they directly impact public perception and understanding of this field. Insufficient public awareness and involvement can result in misunderstandings, apprehension, and skepticism regarding the advantages and dangers of nanotechnology. To promote informed decision-making, ethical reflection, and societal discussion on the societal consequences and ethical issues of nanotechnology, it is necessary to have effective communication, public dialogue, and education efforts.

To tackle these socioeconomic and ethical issues, it is necessary to have collaboration and involvement from several disciplines and stakeholders, including as researchers, policymakers, industry representatives, civil society organizations, and the general public. To fully use the revolutionary capabilities of nanotechnology while ensuring the protection of human health, safety, and welfare, as well as promoting fair and ethical outcomes for society as a whole, it is essential to address these challenges in a proactive manner.

Public perception and education

The societal acceptance, responsible governance, and ethical use of nanotechnology are heavily influenced by public perception and education. These are important factors to consider when it comes to how the general public perceives and understands nanotechnology:

Lack of awareness and comprehension: A significant portion of the general population may possess little information or incorrect beliefs regarding nanotechnology. Education programs focused on enhancing knowledge and enhancing comprehension of the principles, applications, advantages, and risks of nanotechnology are crucial. The outreach initiatives should focus on a wide range of individuals, such as students, educators, lawmakers, industry experts, and the general public.

The ethical and social consequences of nanotechnology should be included in public education, with a focus on promoting critical thinking, ethical reflection, and informed decision-making. Engaging the public in ethical dialogue and decision-making processes requires discussions on the responsible and fair use of nanotechnology, as well as its potential influence on privacy, equity, and human rights.

Risk communication is crucial for addressing public concerns and perceptions related to the possible risks and uncertainties of nanotechnology. Effective and unambiguous dissemination of scientific data, evaluations of potential harm, and regulatory actions plays a crucial role in establishing trust, reliability, and assurance in the safety and management of nanotechnology.

Public Engagement and discourse: Effective public engagement and discourse processes offer stakeholders the chance to express their concerns, viewpoints, and principles about nanotechnology. Participatory methodologies, such as citizen juries, deliberative forums, and stakeholder consultations, facilitate inclusive decision-making processes and promote mutual comprehension and cooperation among a wide range of stakeholders.

Science Communication: Scientists, educators, and communicators have important responsibilities in effectively conveying nanotechnology concepts and discoveries to the public in formats that are easy to understand and interesting. Science communication endeavors should employ a range of channels, such as media outlets, museums, science centers, websites, and social media platforms, to effectively reach various audiences and promote scientific knowledge and public involvement.

Education and Workforce Development: It is recommended that educational programs and curriculum incorporate nanotechnology principles and applications into science, technology, engineering, and mathematics (STEM) education at all levels, ranging from K-12 to higher education and vocational training.

CHAPTER FOUR

Engaging in practical tasks, conducting experiments in a laboratory setting, and participating in projects that involve many academic disciplines improve students' learning and equip them with the necessary skills for pursuing jobs in nanotechnology-related industries.

Community outreach and partnerships are established by collaborative efforts involving academia, industry, government agencies, non-governmental organizations (NGOs), and community organizations to facilitate involvement and outreach to the community. Community-driven endeavors, such as scientific festivals, public talks, and interactive workshops, facilitate conversations, foster trust, and facilitate the sharing of knowledge between scientists and the general public.

Public perception and education activities should prioritize highlighting the significance of ethical conduct, social responsibility, and sustainable practices in nanotechnology research, innovation, and governance. Incorporating ethical issues into research and development processes promotes a culture of responsible innovation and enhances public trust and confidence in nanotechnology.

By promoting public awareness, comprehension, and involvement with nanotechnology, we can enable individuals and communities to actively contribute to shaping its advancement, regulation, and impact on society. Public awareness and education initiatives are crucial for fully harnessing the potential advantages of nanotechnology while effectively tackling its ethical, social, and environmental dilemmas in a responsible and all-encompassing manner.

Prospects for the future and upcoming developments

Nanotechnology's ongoing progress is expected to be influenced by multiple future directions and developing trends, which will have a significant impact on its development and application in various industries. These are the main topics that should be given attention:

Nanomedicine, the merging of nanotechnology and medicine, offers significant potential for individualized healthcare, precise drug administration, and early identification of diseases. Potential future improvements may encompass the creation of intelligent nanosystems with the ability to accurately diagnose, visualize, and cure diseases at the molecular level. Additionally, there is the possibility of combining nanotechnology with regenerative medicine and immunotherapy techniques.

The field of nanoelectronics and quantum computing is driven by the desire to develop electronic devices that are smaller, quicker, and more energy-efficient. Current developments involve the investigation of innovative nanomaterials, such as 2D materials and topological insulators, for the advancement of transistors and quantum bits (qubits) in the future. Quantum computing has the capacity to completely transform computation by allowing for exponential acceleration in solving specific sorts of problems, such as cryptography and optimization.

Nanotechnology is crucial in the development of renewable energy technologies, energy storage systems, and energy-efficient gadgets, as it enables the use of nanomaterials for sustainable energy. Possible future avenues of research could involve the advancement of high-performance solar cells, fuel cells, and batteries through the utilization of nanostructured materials. Additionally, the integration of nanomaterials into energy-efficient lighting, heating, and cooling systems for buildings and transportation could be explored.

Nanotechnology provides novel methods for promoting sustainable agriculture, ensuring food safety, and enhancing nutrition. One of the emerging developments in agriculture is the creation of nanoscale delivery systems for agricultural inputs, like pesticides and fertilizers. These systems aim to enhance effectiveness and reduce the negative effects on the environment. Nanotechnology facilitates the creation of food packaging materials that possess improved barrier qualities, antibacterial capabilities, and sensor functionality for monitoring and preserving food quality.

Nanotechnology is crucial in tackling environmental issues, like pollution, water scarcity, and climate change, by providing effective solutions. Possible future possibilities may involve the advancement of nanomaterial-based strategies for addressing pollution, purifying water, and monitoring air quality. Additionally, nanotechnology might be employed in sustainable manufacturing, recycling, and waste management practices.

The fusion of nanotechnology with life and biotechnology offers novel opportunities in the fields of healthcare, biomanufacturing, and environmental monitoring. Current advancements involve the creation of extremely small biosensors, mechanisms for delivering drugs, and structures for tissue engineering that are used in regenerative medicine. Nanobiotechnology facilitates the manipulation and modification of biomolecules, cells, and tissues for the purposes of diagnostics, therapies, and biomanufacturing.

Nanoscale Manufacturing and 3D Printing: Nanotechnology fosters advancements in manufacturing methods at the nanoscale, additive manufacturing procedures, and tools for nanofabrication. Possible future advancements could involve the advancement of high-resolution 3D printing techniques that can create intricate nanoscale structures with exact manipulation of composition, morphology, and functionality. Nanoscale manufacturing allows for the large-scale production of nanomaterials, nanodevices, and goods with nanotechnology capabilities for a wide range of uses.

Ethical, social, and regulatory factors must be taken into account as nanotechnology progresses to guarantee its responsible and fair development and implementation. Future endeavors may prioritize improving public knowledge and involvement, reinforcing regulatory supervision, and promoting ethical standards in nanotechnology research, innovation, and governance.

Nanotechnology has the capacity to bring about significant breakthroughs in healthcare, electronics, energy, agriculture, and the environment by focusing on future directions and emerging trends. This has the ability to contribute to sustainable development and enhance the overall quality of life worldwide. It is essential to carefully evaluate the societal ramifications, ethical standards, and environmental sustainability when developing and deploying nanotechnology in order to maximize its benefits and minimize hazards.

Conclusion

To summarize, nanotechnology is currently leading the way in scientific and technical advancements, providing exceptional prospects to tackle urgent problems and revolutionize several sectors of society. Nanotechnology has the potential to significantly impact various fields such as healthcare, electronics, energy, and the environment.

Nanotechnology allows for the precise manipulation and control of materials and structures at the nanoscale. This permits the creation of sophisticated materials, devices, and systems that have improved performance, functionality, and efficiency. Applications encompass a wide range of sectors, including as medical, electronics, energy, agriculture, and environmental sustainability.

Nevertheless, the potential benefits of this technology must be balanced with significant concerns regarding safety, ethics, legislation, and societal ramifications. It is crucial to approach the research and implementation of nanotechnology with meticulous evaluation of its possible hazards and advantages, as well as its ethical, social, and environmental consequences.

Public awareness, education, and involvement play a crucial role in promoting informed discussion, ethical contemplation, and responsible decision-making about nanotechnology. Effective collaboration among scientists, policymakers, industry stakeholders, and the public is crucial to ensuring that the benefits of nanotechnology are achieved fairly and in a way that can be maintained over time.

In order to fully harness the positive effects of nanotechnology while mitigating potential hazards and unforeseen outcomes, it is crucial to adhere to values of openness, accountability, and ethical behavior as this field continues to progress.

Nanotechnology will likely have a significant impact on scientific discovery, technological innovation, and societal progress in the future, leading to a more sustainable, equitable, and wealthy future for humanity.

THE END

www.ingramcontent.com/pod-product-compliance
Lightning Source LLC
Chambersburg PA
CBHW070208230526
45471CB00002B/873